AL-ZAHRAWI
(936-1013)
A Pioneer of Modern Surgery

M. BASHEER AHMED, M.D.

Abu al-Qasim al-Zahrawi
(936-1013 CE)
A Pioneer of Modern Surgery

M. Basheer Ahmed, M.D.

ISBN (Print): 978-1-966561-43-9
ISBN (e-Book): 978-1-966561-44-6

Published by
Mary Ethel Eckard
Frisco, TX

This is part of the series of books on Muslim scientists of the medieval era.

CONTENTS

FOREWORD

NAZEER AHMED, PHD

From the ninth to the twelfth centuries, the Islamic world stood at the forefront of human civilization, ushering in an extraordinary era of scientific, medical, and intellectual achievement known as the Islamic Golden Age. During this period, great cities such as Baghdad, Cairo, Cordoba, Samarqand, Nishapur, and Damascus became centers of science, medicine, mathematics, and geography. Libraries, hospitals, and schools supported scholars who were eager to study the natural world and build upon the knowledge of earlier civilizations.

The Abbasid caliphs established the Bait al-Hikma (House of Wisdom) in Baghdad, where scholars from diverse backgrounds came together to translate and study important works from Greek, Persian, and Indian traditions, including the writings of Aristotle, Hippocrates and Galen. They tested the ideas of earlier civilizations, improved them, and added their own discoveries to the human reservoir of knowledge.

In mathematics, Muhammad ibn Musa al-Khwarizmi developed algebra into a clear and systematic field of study. In science, Ibn al-Haytham showed how careful experiments could help us understand light and vision. In medicine, Ibn Sina wrote a medical encyclopedia that was studied in both the Islamic world and Europe for centuries. Travelers such as al Idrisi journeyed across Africa and Asia, recording what they saw and helping people understand distant lands and cultures.

Among these great figures was Abu al-Qasim al-Zahrawi, (936–1013 CE), one of the most important surgeons in history. Living in Muslim Spain, he helped transform surgery into a respected and organized branch of medicine.

His major work, *Kitab al-Tasrif*, was a large medical encyclopedia. The final section focused entirely on surgery and became the most influential surgical manual of the medieval world. What made al-Zahrawi's work special was its clarity. He described operations step by step, explaining which instruments to use, how to perform procedures safely, and how to care for patients afterward. He wrote about treating wounds, setting broken bones, removing bladder stones, assisting childbirth, and performing dental procedures.

He emphasized training, patience, and responsibility in surgical treatment. He was a pioneer in anesthesiology and described ways to reduce pain during operations using herbal mixtures—an early step toward modern practices.

Al-Zahrawi invented, designed and described more than 200 surgical instruments. He carefully illustrated them and explained their uses. By organizing tools and methods so clearly, he made surgery safer and more reliable. Many of his designs look similar to the instruments used in hospitals today.

In the twelfth century, his book was translated into Latin and studied in European medical schools for hundreds of years. His influence spread far beyond his own region and became part of world history.

The achievements of al-Zahrawi and the scholars of the Islamic Golden Age remind us that scientific progress is a shared human achievement. It grows when societies value education, encourage critical thinking, and support scholars and scholarship. Their legacy shows that scientific temper, rational thinking, curiosity, and ethical responsibility can change the world—and that learning from the past can inspire future discoveries.

Dr. Basheer Ahmed, an eminent physician, writes with both authority and insight, bringing the achievements of history's greatest scientists vividly to life. His books, thoughtfully crafted for children, are an invaluable resource that inspires young minds to learn from the past and carry forward the legacy of human discovery.

Nazeer Ahmed, PhD is the Director of American Institute of Islamic History and Culture. Dr. Nazeer Ahmed was a senior scientist on the Moon Landing and Hubble Space Telescope Projects. He is an inventor and holds fourteen United States Patents.

Pages from the Arabic manuscript of the Kitab al-Tasrif.

PREFACE

M BASHEER AHMED

Recognizing Overlooked Contributions

Western historians often highlight the achievements of Greek and Roman scholars up to approximately 300 AD. However, many historical accounts then skip directly to the 16th century and the beginning of the Renaissance, overlooking the 1,300 years in-between, which are commonly referred to as "The Dark Ages of Europe." This viewpoint neglects the significant contributions of Muslim scientists and scholars during this period. In disciplines such as mathematics, science, physics, chemistry, medicine, social sciences, astronomy, and philosophy, Muslim scholars profoundly influenced the advancement of human knowledge, illuminating the world with their discoveries and innovations.

Knowledge and Faith: Foundations for Progress

The Quran emphasizes the importance of both faith and the pursuit of knowledge. This central principle led to the establishment of major institutions, including Bait ul-Hikma (House of Wisdom) and various universities spanning from Baghdad to Cordoba, Spain. Notable Muslim scholars such as Ibn Haitham, Ibn Sina, Al-Razi, Al-Zahrawi, Ibn Rushd, Al-Farabi, Ibn Khaldun, and Ibn Nafis made remarkable contributions across scientific and intellectual fields. Their achievements, inspired by Quranic teachings, laid the foundation for the European Renaissance that emerged in the 16th century.

Decline of Scientific Pursuits

By the 11th century, certain religious leaders and rulers began to discourage the study and teaching of science and philosophy. This trend persisted, and by the 15th century, universities in Muslim countries stopped offering these subjects. As a result, many Muslims—particularly students—are unaware of the vital contributions made by medieval Muslim scholars to world civilizations. Addressing this lack of awareness is a responsibility we all share, ensuring that the achievements of these scholars are acknowledged and celebrated.

Inspiring Future Generations

Sharing this knowledge is essential. The legacy of these scholars should instill pride and inspire young Muslims to become scientists, inventors, and innovators. Learning about the accomplishments of medieval Muslim scholars not only honors their achievements but also motivates the next generation to pursue knowledge and make significant discoveries. This document aims to present the biographies and achievements of these influential Muslim scholars, encouraging future generations to contribute meaningfully to society.

The Dawn of Modern Civilization

Modern civilization began in Baghdad and Muslim Spain, leaving a lasting impact on Europe. European scholars traveled to universities in Baghdad and Cordoba, gaining knowledge that they brought back to their home countries. During this era, the distinguished surgeon Abu Qasim ibn al-Zahrawi was born in Cordoba, Spain.

Abu Qasim ibn al-Zahrawi: Father of Surgery

Al-Zahrawi's greatest achievement is the "Kitab Al-Tasrif," a comprehensive thirty-volume encyclopedia covering medical practice. His work established surgery as a distinct specialty, separate from general medicine and grounded in anatomical knowledge. This earned him the title "father of surgery." "Al-Tasrif" became widely recognized in European universities during the Middle Ages and served as the main reference for surgical studies in institutions in Italy and France.

Beyond his medical and surgical expertise, Al-Zahrawi was adept at using both simple and compound remedies, which also led to his recognition as a "Pharmacist Surgeon."

A Lasting Inspiration

Al-Zahrawi is an inspirational role model for future generations, showing that significant accomplishments require persistence, even in the face of failure. He emphasized the importance of serving humanity above material pursuits. His commitment to surgery was matched by his roles as an author, illustrator, and teacher, underscoring the value of sharing knowledge and mentoring others.

Religious Tolerance and Collaboration in Medieval Spain

During the period in Spanish history from the Muslim Umayyad era in the 700s to the reconquest of Spain in 1492 by the Catholic Monarchs, Muslims, Christians, and Jews lived together in relative peace. This era was characterized by religious tolerance and cooperation among different communities. They jointly conducted research to advance knowledge for the benefit of humanity, producing more scientific approaches that were better equipped to solve human problems.

The Medieval World: A Cradle of Civilization

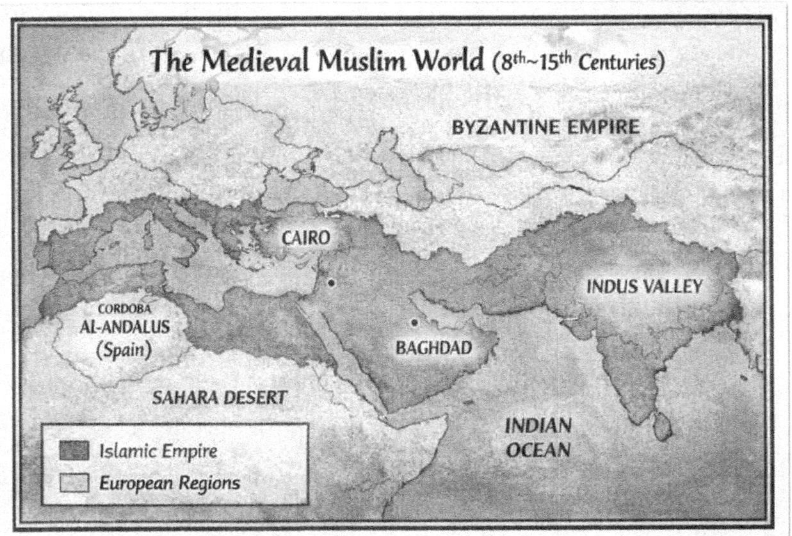

Medieval Muslim World (8th & 15th)

Between the 8th and 15th centuries, much of Europe struggled through a period often called the Dark Ages. During this same time, a brilliant and vibrant civilization was flourishing under Muslim rule, stretching from the Middle East across North Africa and into southern Europe. This civilization became a global center of learning, science, culture, and innovation.

One of its brightest regions was Muslim Spain, known as Al-Andalus. From here, knowledge would eventually spread across Europe and help spark the Renaissance. The city of Cordoba, the capital in the ninth and tenth centuries, was among the greatest cities in the world. It had paved streets, street lighting, public baths, hospitals, and libraries that held hundreds of thousands of books, including priceless works by Aristotle and other ancient scholars—more than all of Europe combined at the time.

Students and scholars from across Europe traveled to cities like Baghdad and Cordoba to study medicine, mathematics, astronomy, philosophy, and science. When they returned home, they carried this knowledge with them, laying the foundation for modern European education and science.

Europeans, in search of seeking knowledge, traveled eagerly to the great universities of Baghdad and Cordoba, returning home to disseminate the knowledge that would later ignite the Renaissance. It was in this extraordinary environment of learning and curiosity that one of the greatest surgeons in history was born.

Cordoba, capital of Al-Andalus, was a city of libraries, hospitals, universities, and advanced urban planning

Al-Zahrawi, the 10th-century father of surgery, emphasized, "Whatever skill I have, I have derived for myself by my long reading of the books of the Ancients and my thirst to understand them until I extracted the knowledge of it from them. I became well-versed in anatomy. Through the whole of my life, I have adhered to experience and practice."

EARLY LIFE OF AL-ZAHRAWI

Abu Qasim Khalaf ibn al-Abbas Al-Zahrawi was born in 936 CE in the town of Al-Zahra, near Cordoba. Known in the West as Albucasis, he would ascend to become not only the court physician to Caliph Al-Hakam II but also the greatest surgeon of the Middle Ages—a figure so foundational that he is rightly called "the father of surgery."

Al-Zahrawi was known not only for his intelligence, but also for his humility, faith, and dedication to helping others. He lived simply and devoted more than fifty years to teaching students, treating patients, and advancing medical knowledge.

The house where he lived and the street named after him (Calle Albucasis) still exist in Cordoba today, preserved by the Spanish government as historical landmarks. These quiet places remind us of a man whose work changed the world.

Al-Zahrawi was trained in the rich academic tradition of Cordoba, where scholars were encouraged to ask questions, observe carefully, and learn from both earlier knowledge and real-life experience. He believed that medicine was not only a science, but also a moral responsibility.

Al Zahrawi's Magnum Opus: Al-Tasrif

Abu al-Qasim Al-Zahrawi (936-1013 CE), physician, surgeon, teacher, and author of Al-Tasrif

Al-Zahrawi's greatest achievement was a massive medical encyclopedia titled: ***Kitab al-Tasrif li-man 'Ajiza 'an al-Ta'lif*** **(*The Arrangement of Medical Knowledge for One Who Is Unable to Compile It*).**

Completed around the year 1000 CE, this work consisted of 30 volumes covering nearly every aspect of medicine known at the time. It included topics such as: General medicine, Pharmacology (drugs and remedies) Dentistry, Childbirth and midwifery and Surgery.

Al-Zahrawi in Apothecary

Because of his deep knowledge of medicines and pharmacology, Al-Zahrawi was skilled in using simple and compound remedies and therefore was also known as the "Pharmacist Surgeon." The translation of Al-Zahrawi's work "Liber Servitors" provides readers with an explanation of how to prepare simple and complex drugs that were generally used in those days. One section of *Al-Tasrif* explains how to prepare simple and complex drugs, many of which were used for centuries.

The final volume—dedicated entirely to surgery—was the most influential. At a time when surgery was often performed by barbers or untrained practitioners, Al-Zahrawi transformed it into a serious scientific discipline.

Al-Zahrawi was, by all accounts, a god-fearing, humble, and pious Muslim. His house and the street where he lived in Cordoba (Calle Albucasis) have been preserved by the Spanish government, standing as quiet testaments to his enduring legacy. His education was rooted in the rich academic traditions of Cordoba, and he devoted over fifty years to teaching, practicing, and revolutionizing medicine and surgery.

Based on a lifetime of meticulous observation and experience, the *Al-Tasrif* covered the full spectrum of medical science, from pharmacology and dentistry to childbirth and surgery. Its most influential section was dedicated to surgery, a field he elevated to unprecedented heights.

Al-Zahrawi using the Mortar and Pestle

The surgical volumes of *Al-Tasrif* were translated into Latin in the 12th century and became the principal reference for surgery in European universities, particularly in Italy and France, during the Middle Ages. For over five hundred years, they served as the standard textbook for surgical education across Italy, France, and other parts of Europe.

Through this influential text, Al-Zahrawi introduced innovative surgical methods and a wide array of instruments, many of which remained in use until the 16th century. The 10th-century European surgeon Pietro Argallata famously declared, "Without doubt, he was the chief of all surgeons." Manuscripts of these translated works are still preserved in libraries in Berlin, London, Paris, and Washington,

D.C. Through *Al-Tasrif,* Al-Zahrawi firmly established surgery as a respected and essential branch of medicine.

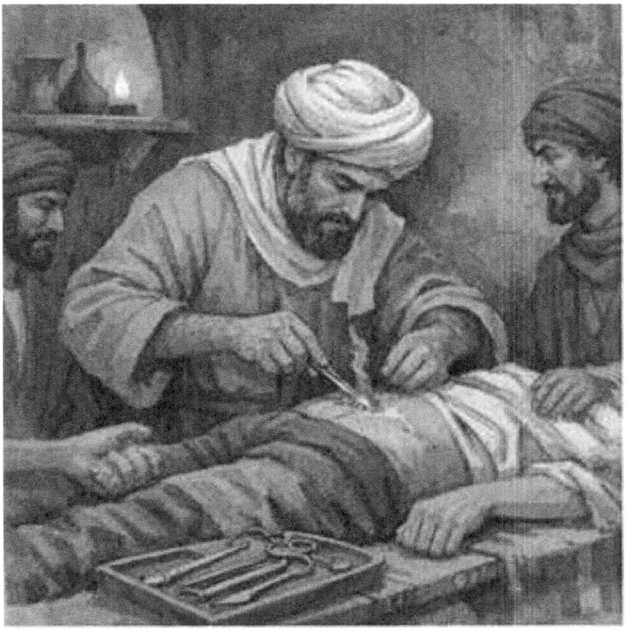

Al-Zahrawi performing an abdominal cauterization

Al-Zahrawi's contributions established surgery as a distinct medical discipline, integrating it into scientific medicine and elevating its status beyond that of barbers or untrained practitioners.

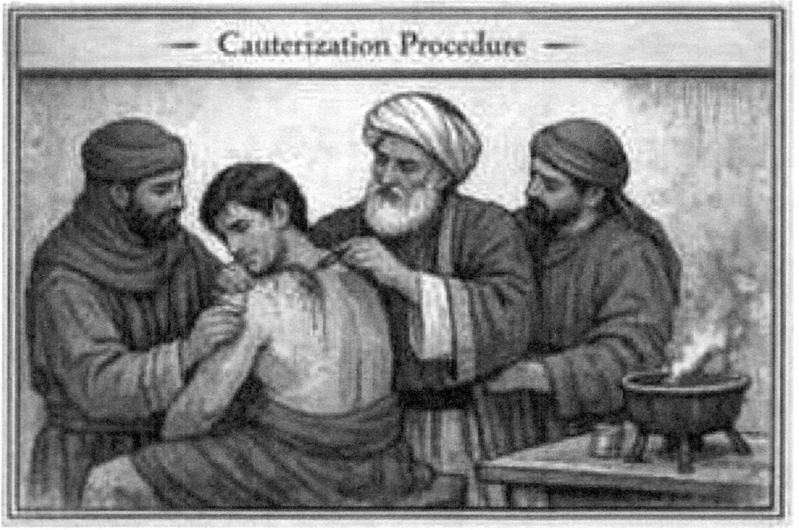

Performing a Cauterization on Shoulder

A Pioneer of Modern Surgery

Al-Zahrawi was a great clinician, surgeon, pharmacist, educationalist, inventor, and illustrator. He understood that surgery's advancement was closely linked to precise anatomical knowledge and specialized tools. "He who devoted himself to surgery," he wrote, "must be versed in the science of anatomy." This conviction was revolutionary at a time when anatomical dissection faced religious and social challenges. Yet, Al-Zahrawi and his contemporaries cultivated a non-dogmatic intellectual atmosphere that encouraged evidence-based inquiry, debate, and the careful study of both animal and human anatomy.

At a time when the study of the human body was controversial, this belief was revolutionary. Yet, Al-Zahrawi and other scholars of this era promoted careful observation, debate, and evidence-based learning.

His surgical inventions, aka instruments, detailed in the final volume of *Al-Tasrif*, numbered over 200. They were not mere adaptations but original creations—scalpels, forceps, retractors, specula, hooks, and bone saws—each designed for a specific and delicate purpose. Many of these tools were so well designed that they remained in use—almost unchanged—until the 16th century. But beyond the instruments themselves, it was his innovative techniques that permanently altered the surgical practice.

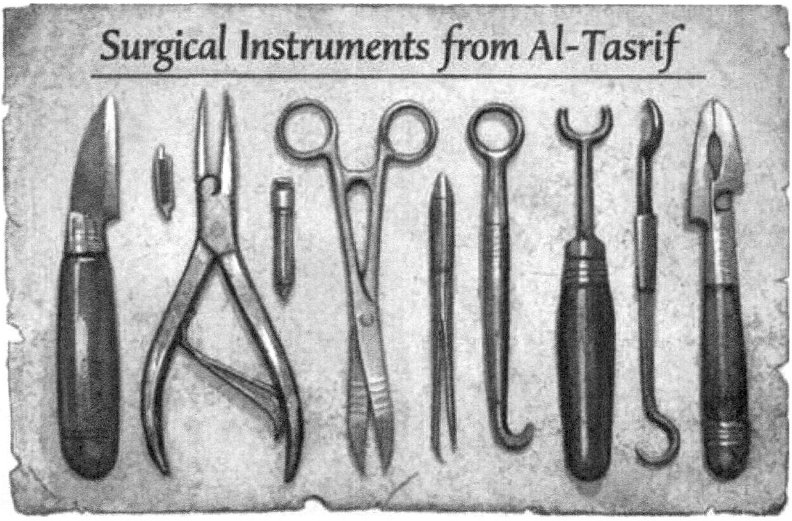

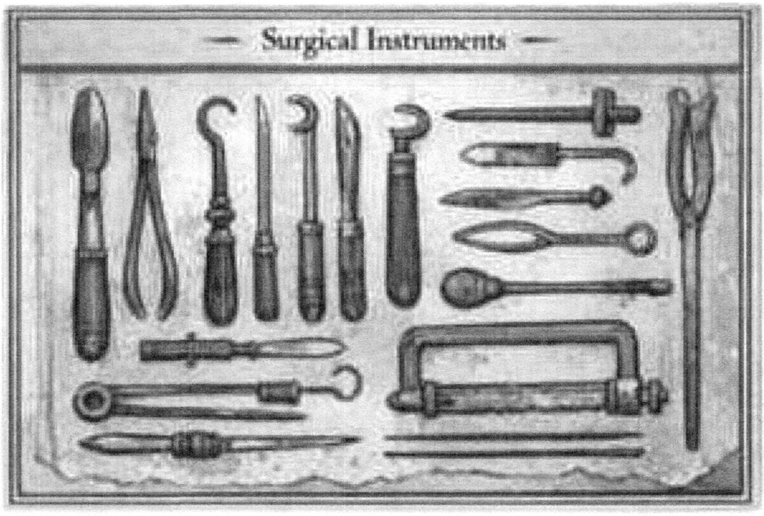

Over 200 surgical instruments designed by Al-Zahrawi, many of which influenced modern tools.

In his 30-volume medical encyclopedia, "Kitab al-Tasrif," Al-Zahrawi stressed that a surgeon must maintain high ethical standards to ensure patient care and successful outcomes. He stated, "A doctor without ethics was a danger to society, no matter how skilled."

Medical Firsts that Shaped the Future

Al-Zahrawi introduced or perfected many techniques still used today:

- He was the first to use catgut sutures, which dissolve naturally inside the body.
- He used cotton for wound dressing and wax to stop bleeding in skull injuries.
- He introduced marking surgical sites with ink, a routine practice today.
- He designed tools to detect and remove bladder stones, laying the groundwork for modern urology.
- He described treatments for varicose veins similar to modern methods.
- He performed tonsillectomies, tracheotomies, and cataract surgeries.

- He identified hemophilia and recognized it as a hereditary disease.
- He warned against dangerous amputations and emphasized patient safety.

He is also considered a pioneer of plastic and reconstructive surgery, as well as an innovator in dentistry, developing techniques for tooth alignment and replacement.

AL-ZAHRAWI THE HEALER: EDUCATION, ETHICS, AND HUMANITY

Al-Zahrawi's legacy extends beyond surgery into the very ethos of medical practice. He established one of the earliest large medical schools in Cordoba, where he insisted on the highest standards from his students—not only in skill, but in ethics and compassion.

He emphasized the sacred doctor-patient relationship. Care, dignity, and respect were central to his medical philosophy, regardless of a patient's wealth or social standing.

Al-Zahrawi was among the first Islamic scholars to advocate for nursing and encourage women to become midwives. His progressive efforts led to significant advancements in hospital and maternal care in Andalusia, as he promoted the training of women as midwives.

This not only improved the quality of maternal care but also ensured that female Muslim patients felt more comfortable receiving treatment, especially when attended by male physicians. Thanks to his dedication, hospitals and nursing care flourished in Andalusia.

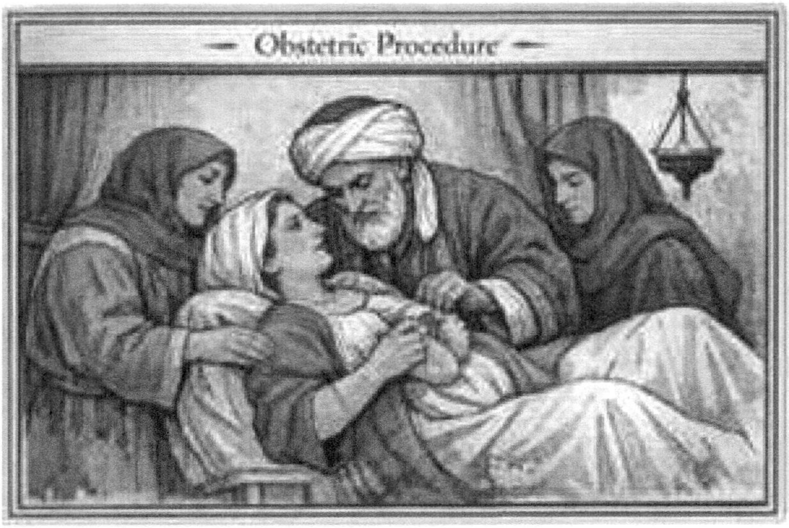

Obstetric Procedure

Humility and Lifelong Learning

Despite his achievements, Al-Zahrawi was, above all, a humble servant of knowledge. His own words in the Al-Tasrif resonate with timeless humility: "Whatever I know, I owe solely to my assiduous reading of the books of the ancients, to my desire to understand them and to appropriate this science; then I have added the observation and experience of my whole life." He saw himself as a link in a chain—a student of the past, a teacher for the future.

Al-Zahrawi Left a Legacy for the New Generation

For centuries, Al-Zahrawi's work, alongside that of Ibn Sina, Al-Razi, and Ibn al-Nafis, formed the canon of world medicine. Yet, a profound intellectual decline in the Muslim world after the 14th century led to the neglect of these very giants. Their works were preserved, studied, and built upon in the West, while fading from memory in the Muslim countries. It was through Latin translations that his monumental work on surgery was kept alive, illuminating the path for figures like Guy de Chauliac, who quoted him over 200 times in the 14th century.

Today, as we witness a renewed glimmer of hope for the scientific pursuit in the Muslim world, Al-Zahrawi's legacy offers more than historical pride; it provides a powerful roadmap forward.

To young students, his story is not just history, it is a call to action. He teaches us that faith and reason are not enemies, and that seeking knowledge is an act of worship.

Al-Zahrawi demonstrated intellectual courage by questioning, experimenting, and documenting in an age of limited tools. He faced obstacles and opposition yet responded with brilliant innovation rather than resignation. His life proves that perseverance is what truly unlocks human potential.

No matter your chosen field—medicine, engineering, science, art, or leadership—you can follow his example: Serve humanity, Seek truth, Act ethically, Stay humble, Keep learning.

"Only by repeated visits to the patient's bedside can the physician follow the progress of his medical treatment." Al-Zahrawi

A Final Reflection

Civilizations that abandon the pursuit of knowledge surrender their future. You have the duty to revive the spirit of inquiry. Keep the example of the great Muslim scholars of Baghdad and Cordoba before you. The frontiers may have changed—quantum computing, genetic engineering, climate science, AI—but the required spirit remains: audacious inquiry, ethical rigor, and a commitment to human service and universal betterment.

Let Al-Zahrawi's humility guide you. He attributed his wisdom to his teachers and to his own diligent experience. Be a perpetual student of the universe. His legacy proves that when you dedicate your skills to alleviating human suffering and illuminating the darkness of ignorance, you achieve a form of immortality.

You are the descendants of the builders of civilizations, the children of pioneers who mapped the stars, named the stars, and healed broken bodies with exceptional, unparalleled skill. The world needs both your intellect and your faith. Honor Al-Zahrawi's legacy through study, innovation, and services, so you may inspire and illuminate the path for generations to come.

Timeline: Life and Legacy of Al-Zahrawi

936 CE

- Birth of Abu al-Qasim Al-Zahrawi in Al-Zahra, near Cordoba, Spain

10th Century

- Cordoba flourishes as a global center of learning
- Al-Zahrawi studies medicine, anatomy, pharmacology, and surgery
- Begins medical practice and teaching

Circa 1000 CE

- Completion of Al-Tasrif, a 30-volume medical encyclopedia
- Final volume dedicated entirely to surgery

11th–12th Centuries

- *Al-Tasrif* translated into Latin
- Adopted as the main surgical textbook in European universities

12th–16th Centuries

- Al-Zahrawi's surgical instruments and techniques used across Europe
- Surgery recognized as a scientific medical discipline

14th Century

- Surgeon Guy de Chauliac frequently cites Al-Zahrawi in his own work

Modern Era

- Al-Zahrawi recognized worldwide as the Father of Modern Surgery
- His legacy studied in medicine, ethics, and history of science

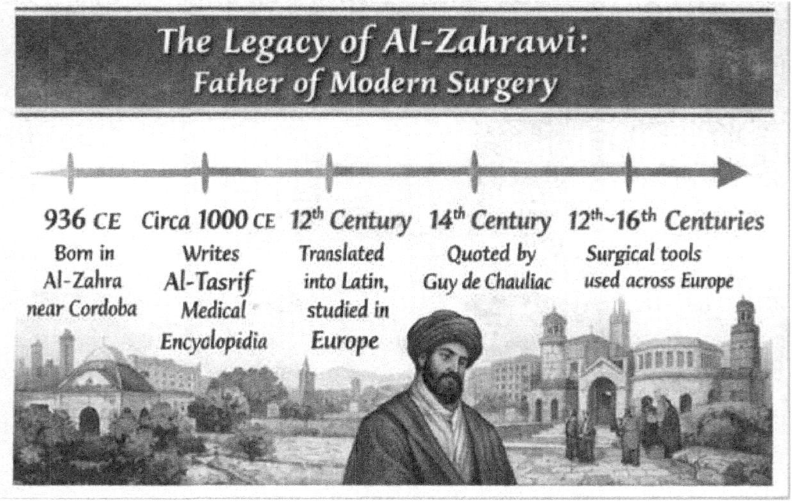

Latin translations of Al-Zahrawi's work shaped European medical education for over 500 years

The postage stamp issued in honor of Al-Zahrawi commemorates one of the most influential physicians and surgeons of the Islamic Golden Age. Known in the West as Abulcasis, Al-Zahrawi is celebrated for his groundbreaking contributions to surgical techniques and medical knowledge. It symbolizes the importance of knowledge, innovation, and cross-cultural contributions to the advancement of healthcare.

BIBLIOGRAPHY

1. Levey M., Early Arabic Pharmacology, E. J. Brill, Leiden, 1973
2. Jones, PM. "Medieval medical miniatures." London: The British Library, in association with The Wellcome Institute for the History of Medicine; 1984 - pages 27-29
3. Savage-Smith E. "Attitude toward dissection in medieval Islam." J Hist Med Allied Sci. 1995550: 67-110
4. Hehmeyer, I. & Khan, A. "Islam's Forgotten Contribution to Medical Science." Canadian Medical Association Journal, page 176, 2007.
5. Gimovsky, M.L., Soo Han, J. "Reducing the medico legal risk of vacuum extraction." OBG management, Volume 19 No. 6, June 2007, page 84.
6. Lunde, Paul. "Science in Al Andalus," Published in Science: The Islamic Legacy. Aramco-Washington D.C. 1988, page 21-26.
7. Azmi, A.A., "Contribution of Muslim Physicians in the Middle Ages," Hamdard Vol. XXIX 1-2 ig SS-Hamdard Publications, Karachi, Pakistan.

8. Ashoor, A.Z. "Muslim Medical Scholars and their Work," Islamic World Medical Journal. January 1984, pages 49-50.
9. Al Ghazal S. K., "Zahrawi and Plastic Surgery," ArabMed Journal, Issue 2 (12); 2002.
10. Hayek, Simon. "How Al- Zahrawi Reached the Occident," Islamic World Medical Journal. February 1984, page49-52.
11. Hamarneh, Sami K. "Vistas of Arabic Healing Arts in Theory and Practice," Hamdard XXXII No. 3, September 1989, pages 3-54; 16-18
12. Huff, T.E. "The Rise of Early Modern Science: Islam, China, and the West," Cambridge University Press, 2003

STUDENT QUIZ

A. Why was Cordoba considered one of the greatest centers of learning in the medieval world?
B. How did Muslim Spain (Al-Andalus) help preserve and spread knowledge to Europe?
C. Why do you think European students traveled to cities like Cordoba and Baghdad to study?

Science and Innovation

D. What made Al-Zahrawi's approach to surgery different from other medical practices of his time?
E. Why was Al-Zahrawi's emphasis on anatomy considered revolutionary?
F. Which of Al-Zahrawi's medical inventions or techniques do you find most impressive, and why?

Ethics and Humanity

G. How did Al-Zahrawi view the doctor-patient relationship?
H. Why was his support for nursing and women midwives important for society?

I. In what ways did Al-Zahrawi combine scientific skill with compassion and ethics?

Faith and Reason

J. How did Al-Zahrawi show that faith and scientific inquiry can work together?
K. Why did he consider the pursuit of knowledge a form of worship?

Lessons for Today

L. Why do civilizations decline when they stop valuing education and science?
M. How can Al-Zahrawi's life inspire students today, even outside the field of medicine?
N. What qualities—besides intelligence—helped make Al-Zahrawi successful?
O. How can young people today revive the spirit of inquiry shown by scholars of Cordoba and Baghdad?

QUESTIONS

Part 1: Multiple Choice (See the Answer Key at the end of this section)

Where was Al-Zahrawi born?

A. Baghdad
B. Cairo
C. Al-Zahra near Cordoba
D. Damascus

What was the name of Al-Zahrawi's major medical encyclopedia?

A. Canon of Medicine
B. Al-Tasrif
C. Book of Healing
D. Kitab al-Qanun

How many volumes were in Al-Tasrif?

A. 10
B. 20
C. 30
D. 40

Which field did Al-Zahrawi most strongly transform into a scientific discipline?

A. Pharmacy
B. Astronomy
C. Surgery
D. Chemistry

Why was Al-Zahrawi's work important to Europe?

A. It was written in Greek
B. It replaced Roman medicine
C. It became a standard surgical textbook for centuries
D. It focused on philosophy

Part 2: True or False

True / False: Al-Zahrawi believed surgery could be practiced without knowledge of anatomy.

True / False: He introduced dissolvable internal sutures made from catgut.

True / False: Al-Zahrawi encouraged women to work as midwives and nurses.

True / False: His surgical instruments were copied and used in Europe for over 500 years.

True / False: Al-Zahrawi believed knowledge should only be kept within the Muslim world.

Part 3: Short Answer

Why was Cordoba considered a center of learning in the medieval world?

Name two medical innovations introduced by Al-Zahrawi.

How did Al-Zahrawi combine ethics with medicine?

Part 4: Reflection Questions

What lesson can modern students learn from Al-Zahrawi's life?

Why is Al-Zahrawi called the "Father of Modern Surgery"?

ANSWERS

PART 1: MULTIPLE CHOICE (ANSWER KEY)

C – Al Zahrawi was born in Al-Zahra near Cordoba

B – Al-Zahrawi's major medical encyclopedia was *Al-Tasrif*

C – There are 30 volumes in *Al-Tasrif*

C – Al-Zahrawi most strongly transformed surgery

C – Al Zahrawi's work was important to Europe because it became a surgical textbook for centuries

PART 2: TRUE OR FALSE (ANSWER KEY)

False

True

True

True

False

PART 3: SHORT ANSWER (ANSWER KEY)

Cordoba was considered a center of learning in the medieval world because it had universities, hospitals, libraries with hundreds of thousands of books, and scholars from many regions studying science, medicine, and philosophy.

Any 2 of the following medical innovations that were introduced by Al-Zahrawi

Catgut sutures
Surgical instruments
Ink marking before surgery
Bladder stone removal tools
Cotton dressings

Al-Zahrawi combined ethics with medicine by emphasizing patient dignity, compassion, safety, and moral responsibility in medical practice.

PART 4. REFLECTIVE QUESTIONS (ANSWER KEY)

Sample Answer: Students can learn the importance of curiosity, lifelong learning, humility, ethical service, and the balance between faith and reason.

Because he established surgery as a scientific discipline, designed original surgical instruments, and wrote textbooks used for centuries.

Scoring Guide

13–15 correct answers: Excellent understanding

10–12 correct answers: Very good

7–9 correct answers: Good, review recommended

Below 7 correct answers: Needs revision

Made in the USA
Middletown, DE
31 October 2026

36002851R10030